체인지업그라운드 포항

포스코 A&C

운생동건축 장윤규

FOREWORD

더불어 함께 발전하는 기업시민

포스코그룹의 백년대계를 마음에 새기며 경영이념 <더불어 함께 발전하는 기업시민>을 선포한지도 어느새 4년이 되었습니다. 그동안 기업시민으로서 포스코는 스스로가 사회 구성원의 일원이 되어 임직원, 주주, 고객, 공급사, 협력사, 지역사회 등 여러 이해관계자와 더불어 함께 발전하고, 배려와 공존, 공생의 가치를 함께 추구하는데 앞장서 왔습니다.

특히 1조원 규모의 펀드를 조성하고, '또 하나의 포스코'를 육성하기 위해 벤처기업 인큐베이팅센터인 'CHANGeUP GROUND(체인지업그라운드)'를 지난해 포항에 건립하는 등 포스코그룹은 청년 일자리 창출과 지역 경제 활성화에 있어서도 늘 재계의 모범이 되어 왔습니다.

체인지업그라운드는 한국어로는 "창업(創業) 그라운드", 영어로는 "CHANGeUP GROUND"로 <창업>과 <혁신>이라는 두가지 의미를 함축적으로 담았으며, 포스코그룹은 체인지업그라운드 포항을 청년 창업가들에 대한 지원과 해외 진출이 가능한 최고의 Playground로 만들기 위해 투자를 아끼지 않고 있습니다.

체인지업그라운드 포항은 세계 최고 수준의 연구 중심 대학 '포스텍', 산업기술 및 소재분야 연구기관 포항산업과학연구원(RIST) 등과 함께 세계 유일의 산학연 협력 체계를 구축하고 있으며, 체인지업그라운드 기반의 포항은 미국의 '실리콘 밸리'에 이은 대한민국의 '퍼시픽 밸리'로 진화하고 있습니다. 지난 6월말 현재, 체인지업그라운드 포항에는 스타트업 87개사가 입주해있으며 이들 기업의 기업가치는 약 10,177억원에 달합니다.

장윤규 건축가님께 체인지업그라운드를 '누구에게나 열려있는 우주선 컨셉의 건물'로 설계해 달라는 당부을 드린 기억이 납니다. 이 곳에 오는 청년들이 일반 사무실같이 정형화되거나 좁아서 움직임에 제한이 있는 공간이 아닌, 무한한 상상력을 발휘함으로써 구글이나 애플같은 혁신기업 탄생의 첫 걸음이 될 수 있는 창의적인 공간으로 조성하고 싶다는 뜻이었습니다.

체인지업그라운드는 세계적인 기업으로 성장을 꿈꾸는 유망 스타트업들의 요람입니다. 53년전 영일만 황무지에서 자본도 경험도 모든 것이 부족했지만 오늘의 포스코에 이른 것처럼, 체인지업그라운드가 미래를 이끌어갈 기업들이 이 곳에서 탄생하는 대한민국의 벤처 생태계 중심이 됨은 물론, 스티브 잡스, 빌 게이츠, 마크 저커버그와 비견되는 기업가들의 사관학교가 되길 기대합니다.

창업의 희망, 벤처밸리의 꿈을 꿀 수 있게 지원해주시고, 상상력을 자극하는 아름다운 건물을 세상에 선보이기 위해 애써주신 운생동 건축사무소 및 시공테크, 포스코홀딩스 산학연협력담당, 포스코 투자엔지니어링실, 포스코건설, 포스코A&C 등 모든 관계자들의 헌신적인 노력에 감사드립니다.

체인지업그라운드 포항을 위해 수고해주신 모든 분들의 헌신과 노력은 대한민국 청년들에게 큰 힘과 격려가 될 것이며, 그들이 만들어 갈 위대한 대한민국의 미래에 든든한 밑거름이 될 것입니다.

2022년 7월

최정우, 포스코홀딩스 회장

Corporate Citizenship: Building a Better Future Together

Four years have passed since we set out a far-sighted plan for POSCO Group and declared the management philosophy of <Corporate Citizenship: Building a Better Future Together>". In that time, we at POSCO have taken the lead in development and grown together with our stakeholders — our employees, stockholders, customers, suppliers, partners, and the community — by becoming a member of society as a corporate citizen and by pursuing the values of consideration, coexistence, and symbiosis.

POSCO Group has always been a model for the business world in creating jobs for youth and revitalizing the local economy. Indeed, last year we created a KRW 1 trillion venture fund and built 'CHANGeUP GROUND', an incubating center for venture companies, in Pohang to nurture 'another POSCO'.

CHANGeUP GROUND, "창업(創業) 그라운드 (pronounced Chang-up Ground, meaning Start-up Ground)" in Korean, has both the meanings of <start-up> and <innovation>, and POSCO Group is sparing no expense in making CHANGeUP GROUND into the ideal playground for supporting young entrepreneurs and helping them to expand overseas.

CHANGeUP GROUND is establishing the world's only industry-academic-research collaboration system with 'POSTECH', the world's top research-oriented university, and the Pohang Research Institute of Industrial Science & Technology (RIST), the industrial technology and material research institute, and because of this, Pohang is evolving into Korea's Pacific Valley following the model of Silicon Valley in the U.S. As of the end of June, 87 start-up companies have moved into CHANGeUP GROUND, and the enterprise value of these companies is approximately KRW 1.177 trillion.

I remember asking architect Yoon-Gyoo Jang to design CHANGeUP GROUND as a 'spaceship concept building that is open to everyone'. My hope was to create, not a standardized space like a regular office or a narrow space that limits the motions of the young people who come here, but a creative space that allows them to demonstrate unlimited imagination and a place where they can take the first step in the creation of innovative companies like Google or Apple.

CHANGeUP GROUND is the birthplace of promising start-ups that dream of growing into global companies. Just as POSCO started in the wastelands of Yeongil Bay 53 years ago, lacking both capital and experience, and yet became what it is today, I hope that CHANGeUP GROUND can become the center of Korea's venture ecosystem, where companies that will lead the future are born, and an academy for entrepreneurs comparable to Steve Jobs, Bill Gates, and Mark Zuckerberg.

I would like to extend my gratitude to all stakeholders, including the UNSANGDONG Architects Cooperation, SIGONG Tech, POSCO Holdings Industry-Academic-Research Cooperation Manager, POSCO Investment Planning & Engineering Office, POSCO E&C, and POSCO A&C, for their dedicated efforts, as well as their support for the dream of a Venture Valley, a hope for start-ups, and their hard work to present a stunning building that stirs the imagination.

The dedication and efforts of all those who worked hard for the CHANGeUP GROUND will be a great source of strength and encouragement for young Koreans and will become a solid foundation for the great future of the Republic of Korea they will create

July 2022

Choi, Jeong-Woo, CEO, POSCO HOLDINGS

또 하나의 퍼시픽 밸리, 체인지업그라운드 포항

포스코그룹의 벤처 인큐베이팅 센터인 체인지업그라운드 포항이 꿈꾸고 있는 '또 하나의 퍼시픽 밸리(Another Pacific Valley)'의 의의와 비전은 아래와 같습니다. 포스코는 68년 제철보국의 경영이념으로 시작하여 오늘날 세계 최고의 철강 기업으로 우뚝 섰고 이제는 철강을 넘어 친환경 미래소재 대표기업으로 거듭나고 있습니다. 포스코는 사업으로 창출한 이익을 바탕으로 국내 역사상 최대인 2조 원의 기부를 통하여 교육보국의 철학을 가진 포스텍과 RIST 그리고 가속기연구소 등 세계적인 연구시스템을 구축하였습니다.

2018년 취임한 최정우 회장님은 제철보국, 교육보국을 아우르며 기업 스스로가 사회 구성원의 일원이 되어 임직원, 주주, 고객, 공급사, 협력사, 지역사회 등 여러 이해관계자와 더불어 함께 발전하고, 배려와 공존, 공생의 가치를 함께 추구해 나가야 한다며 '기업시민'을 새로운 경영이념으로 선포하셨습니다. 이러한 기업시민 경영이념을 바탕으로 포스코그룹은 공존과 공생의 가치를 실현하고자, 국가 연구결과를 상용화하기 위한 1조 원 벤처펀드를 조성하고 벤처기업의 안정적인 성장을 지원하고자 체인지업그라운드를 설립하는 등 혁신보국의 범포스코 벤처생태계도 구축해 나가고 있습니다. 체인지업그라운드는 포스코그룹의 기업시민 5대 브랜드 중 하나인 '함께 성장하고 싶은 회사(Challenge with POSCO)'를 구현하여 국가 발전에 기여하겠다는 철학을 바탕으로 하고 있습니다. 포스코가 한 기업으로서의 이익 창출과 성장에 멈추는 것이 아니라, 벤처기업의 성장 지원을 통해 국가 차원의 일자리 창출과 경제 활력 제고에도 기여하여 더 나은 미래를 만드는 데 일조해야 한다는 것입니다.

포스코 그룹은 80개국에 162개의 법인과 75조 원의 매출을 하는 제계 6위의 비즈니스 시스템을 가지고 있고, 또한 가속기 2대를 포함하여 단일캠퍼스로는 세계 2위에 해당하는 2조 원의 연구시설과 5,000명의 연구원이 매년 1조원의 연구비를 사용하는 연구 시스템도 가지고 있습니다. 그리고 경상북도와 포항시 그리고 전라남도와 광양시 등 강력한 지자체지원 시스템을 구축하였습니다. 또한 1조 원 펀드를 기반으로 900개가 넘는 벤처생태계를 구축하고 있습니다. 이러한 비즈니스시스템, 연구시스템, 지자체 지원시스템 및 벤처생태계가 체인지업그라운드에서 모든 자원을 융합하여 청년창업가들에게 세계 최고의 플레이그라운드를 제공한다는 의의가 있습니다. 이를 통하여 창업자들이 혁신을 이루어서 전세계로 뻗어갈 수 있을 것으로 예상됩니다.

이런 의의를 바탕으로 포스코그룹은 850억 원을 출연하여 19개월 동안 28,000 평방미터의 비수도권 최대 규모의 창업공간인 체인지업그라운드 포항을 완공하였습니다. 이 체인지업그라운드 포항은 현재 산학연관 협력으로 87개의 벤처기업이 입주하여 1조 원의 시가총액과 801명의 종업원들이 근무하고 있습니다. 수도권에서 12개의 기업이 본사를 포항으로 이전하였고 5개의 기업이 포항 사무실을 개소하여 75명의 종업원을 채용하는 등 지방경제 활성화와 청년일자리 창출에 기여하고 있습니다.

본 체인지업그라운드 포항은 단순히 일반적인 창업 인큐베이팅센터가 아닙니다. 포스코그룹의 연구시스템이 진행하고 있는 소부장, 바이오, IT 분야의 최첨단 연구결과가 상용화 될 수 있도록 각 분야별 인큐베이팅이 가능한 세계 최초의 인큐베이팅 콤플렉스 개념을 도입하였습니다. 최첨단 소부장 기술이 있지만 공장 가동 경험이 없는 소부장 벤처를 위하여 중기부와 함께 제조전용 인큐베이팅센터를 구축하여 포스코의 공장가동 노하우를 전수하고 있습니다. 바이오 분야는 실험이 가능한 WetLab 기반의 바이오전용 인큐베이팅센터를 통하여 30여 개의 벤처기업을 유치하고 있습니다. 또한 경상북도 및 포항시와 협력하여 경상북도 전체 관공소 및 교육기관, 그리고 가속기연구소와 벤처기업들이 활용하는 데이터센터를 기획하고 있습니다. 본 데이터센터는 양자컴퓨터가 장착된 최첨단 데이터센터가 될 예정입니다.

체인지업그라운드 포항은 산학연관 협력을 통하여 포스코그룹의 모든 자원을 활용하고 청년창업자들에게 세계 최고의 플레이그라운드를 제공하여 포스코보다 더 큰 사업들을 창출하고 지방이 소멸되는 위기에 지방경제 활성화에 기여할 것입니다. 청년들에게 양질의 일자리를 제공하며 포스코를 넘어서 국가적인 문제를 해결하는 기업시민으로서 역할을 수행하여 태평양 동쪽의 실리콘밸리에 비견되는 또 하나의 퍼시픽 밸리를 꿈꾸는 것입니다..

2022년 7월

박성진, 포스코홀딩스 산학연협력담당 전무

Another Pacific Valley, CHANGeUP GROUND

The basis for the vision of POSCO Group's venture incubation center CHANGeUP GROUND as "Another Pacific Valley" is as follows: In 1968, POSCO started out with a management philosophy of producing steel to support economic development. Today, it stands as the world's best steel company. But moreover, it goes beyond steel, reborn as one of the leading companies in ecofriendly future materials. Based on the profits generated through its business and based on the philosophy that patriotism can be fostered through education, POSCO has established a world-class research system that includes POSTECH, RIST, and Pohang Accelerator Laboratory, through a donation of KRW 2 trillion, the largest in domestic history.

CEO Choi, Jeong-Woo, who took office in 2018, says that the company itself should become a member of society, encompassing steelwork and education as patriotism, and grow together with stakeholders including employees, shareholders, customers, suppliers, business partners, and local communities. He declared "Corporate Citizenship" as the Group's new management philosophy, saying that they should pursue the values of kindness, coexistence, and symbiosis. Based on this corporate citizenship management philosophy, POSCO Group created a KRW 1 trillion venture fund to commercialize national research results to realize the value of coexistence, and established CHANGeUP GROUND to support the stable growth of venture companies. It is also building a pan-POSCO venture ecosystem. CHANGeUP GROUND is based on the philosophy of contributing to national development by realizing "Challenge with POSCO", one of the five corporate citizenship brands of the POSCO Group. Instead of merely generating profits and growing as a company, POSCO strives to do much for the making of a better future by contributing to job creation and economic vitality at the national level by supporting the growth of venture companies.

POSCO Group has 162 corporations in 80 countries and has the 6th-largest business system, with sales of KRW 75 trillion. In addition, it has a research system worth KRW 2 trillion, which is the second largest in the world for a single campus including two accelerators, and a research system in which 5,000 researchers spend a KRW 1 trillion in research funds every year. In addition, there is strong local government support systems in Pohang-si (Gyeongsangbuk-do) and Gwangyang-si (Jeollanamdo).

It is currently building a venture ecosystem involved in 900 ventures with a KRW 1 trillion fund, providing young entrepreneurs with the world's best playground, a convergence of resources from the business system, research system, local government support system, and venture ecosystem in CHANGeUP GROUND. The expectation is that founders will be able to innovate and expand through this all over the world.

Based on this significance, POSCO Group contributed KRW 83 billion to complete CHANGeUP GROUND. Its 28,000 square meters was completed in 19 months and now stands as the largest start-up space in the non-metropolitan area.

Currently, 87 venture companies have moved into CHANGeUP GROUND through industry-academia cooperation, with a market capitalization of a KRW 1.177 trillion. and 801 employees. Twelve companies moved their headquarters from the metropolitan area to Pohang, and five companies opened an office in Pohang and hired 75 employees, contributing to revitalization of the local economy and job creation for youth.

CHANGeUP GROUND is not your typical startup incubation center. They introduced the concept of the world's first incubating complex that enables incubation across fields so that state-of-the-art research results in materials/machine parts/equipment, bio, and IT, conducted by POSCO Group's research system, can be commercialized. For materials/machine parts/equipment ventures that have cutting-edge technology but no factory operation experience, POSCO established an incubating center dedicated to manufacturing together with the Ministry of SMEs and Startups to pass on POSCO's factory operation knowledge. The bio sector is attracting about 30 venture companies through the WetLab based bio-dedicated incubating center, a prime place for conducting experiments . In addition, in cooperation with Gyeongsangbuk-do and Pohang-si, they are planning a data center used by all government offices and educational institutions in Gyeongsangbuk-do, as well as accelerator research institutes and venture companies. The data center will be a state-of-the-art data center equipped with quantum computers.

CHANGeUP GROUND uses all the resources of the POSCO Group through industry-academia cooperation, to provide young entrepreneurs with the world's best playground. Businesses larger than POSCO will be created, contributing to the vitalization of the local economy during a time when the provinces face economic extinction, and high-quality jobs are provided to young people. Its dream is to go beyond POSCO to serve as a corporate citizen that solves national problems and becomes a Pacific Valley, comparable to Silicon Valley

July 2022
Park, Seong Jin, Executive Vice President,
Industry-Academy-Research Cooperation, POSCO HOLDINGS

CONTENTS

CONCEPT

소통의 플랫폼 - 체인지업그라운드 포항
Platform of Communication - CHANGeUP GROUND

미래산업을 준비하는 연구중심의 공간으로 포항공대에 계획되는 창업지원센터인 체인지업그라운드 포항은 스타트업 벤처기업의 육성을 위한 중심거점이다. 다양한 분야의 연구들이 교류하고 소통하는 공간으로 청년 창업 프로그램의 중심이 되어야 한다. 창업지원센터의 공간은 벤처의 박스들이 모여서 하나가 되는 플랫폼을 만들어내는 건축이어야 한다. 창업지원센터의 플랫폼은 열려 있으며 서로간에 넘나들며 융합되는 방식을 제안한다.

CHANGeUP GROUND, a startup support center to be built at Pohang University as a research-oriented space to prepare for future industries, is a central base for nurturing start-up venture companies. It is to be the center of the youth entrepreneurship program as a space where researchers in various fields can exchange and communicate. The space of the Startup Support Center should be an architecture that creates a platform where the boxes of ventures come together to become one. The platform of the Startup Support Center is open and suggests a method of convergence that crosses each other.

미래등대 공장

체인지업그라운드 포항은 포스코의 더불어 함께 발전하는 기업정신을 담아내는 소통의 공간이기도 하다. 미래를 밝게 비추고 새로운 세상을 향한 젊은 창업인들을 위한 등대와 같은 건축이어야 한다. 창업지원센터는 미지의 우주를 향해 항해를 떠나는 거대한 우주선과 같은 공간이다. 철로 이루어진 거대한 매스를 마치 떠있는 건축으로 구현함으로서 미래에 대해 비상하는 건축적 이미지를 구현하였다.

Future Lighthouse Factory

CHANGeUP GROUND is also a space for communication that captures the corporate spirit that develops together with POSCO. It should be an architecture that illuminates the future and is like a lighthouse for young entrepreneurs who are heading towards a new world. The Startup Support Center is a space like a giant spaceship sailing toward an unknown space. By realizing a huge mass of iron as a floating architecture, an architectural image soaring toward the future is realized.

플로팅 건축

미래에 대한 상상을 꿈꾸는 공간으로 가능성으로 열려진 자유도를 획득하는 건축이여야 한다. 떠있는 건축으로서 자유도를 획득하는 플로팅(floating)의 개념을 적용하려 한다. 무중력적인 사고를 통하여 가벼움의 철학을 통해 달성되어질 수 있는 다양한 다양성을 구축하였다. 벤처를 이루는 기본적인 유니트의 박스를 중간 플랫폼의 매스로 구성하고 저층부는 무중력적으로 쌓아 올린 박스의 조합을 상층부에는 하층부의 매스조합을 미러시켜 쌓아올린 박스의 조합을 구성한다.

플로팅에 의해 열려진 건축은 물리적 실체로부터 자유롭고 대지를 둘러싼 포항공대의 타 연구소와의 도시적 관계를 연결해주는 강력한 도구로 작용한다. 단순히 도시적 컨텍스트가 비슷한 매스의 조합에 이루어지기를 원한다면, 미래적 컨텍스트는 도시적 역할에 의해서 만들어지는 새로운 건축적 형식을 요구한다고 생각하였다. 중간 매개체적 그라운드레벨의 자유도는 자연스럽게 학제간의 소통을 강화하고 끌어드리는 장치로 작용한다. 도시적 흐름에 대한 대응이다. 가장 도시적인 건축은 흐름을 막지 않고 받아들이는 관계의 설정이다.

Floating Architecture

It should be an architecture that acquires degrees of freedom that is open to possibilities as a space to dream of the future. As a hovering architecture, we try to apply the concept of floating, which acquires degrees of freedom. Through weightless thinking, diversity that can be achieved through the philosophy of lightness was established. A box, which is the basic unit that makes up the venture, composes the mass of the middle platform; the lower part is a combination of weightless stacked boxes, and the upper part forms a combination of stacked boxes by mirroring the mass combination of the lower part.

The architecture opened by floating is free from physical reality and acts as a powerful tool to connect the urban relationship with other research institutes of POSTECH surrounding the site. If we simply want this to be achieved through a combination of masses with a similar urban context, we thought that the futuristic context requires a new architectural form created by the urban role. The degree of freedom of the ground level as an intermediate medium naturally acts as a device to strengthen and attract interdisciplinary communication. It is a response to the urban trend. The most urban architecture is the setting of a relationship that embraces the flow without blocking it.

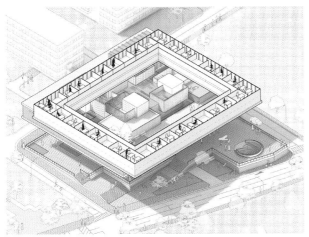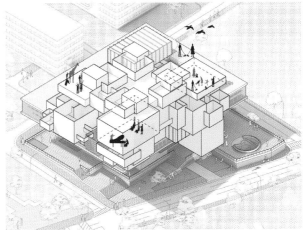

신화적 상상력

떠있다는 사실을 통하여 몽상적이며 신화적인 세계로 끌어들인다. 대지로부터 떨어져 부유한다. 도시와
건축은 이제 단순히 대지로부터의 무게를 지탱하는 도구가 아니라 물리학적 중력을 지배하는 주체자가 된다.
플로팅에 의해 비워진 대지는 활동성, 풍부함, 연결성 등으로 변환된다. 단순히 기능공간으로서가 아니라
수많은 도시적 요소들을 받아들이는 일종의 가속장치가 된다. 칭입지원센터는 이러한 도시직 사회적 정보들을
획득하는 장치로 건축을 변화 시키는 새로운 유전자로서의 건축이며 새로운 상상력의 신화가 되어야 한다.

Mythical Imagination

Through the fact that it is floating, it draws the beholder into a dreamy and mythical world. It
floats, separated from the ground. Cities and architecture are no longer simply tools that support
the weight above the ground, but become the subjects that dominate the physical gravity. The
land emptied through floating is transformed into activity, abundance, and connectivity. It is
not simply a functional space, but a kind of accelerator that accepts numerous urban elements.
The start-up support center is a device that acquires such urban and social information; it is
architecture as a new gene that changes architecture, and it should become a myth of new
imagination.

PROGRAM

인큐베이팅 용광로

An Incubating Furnace

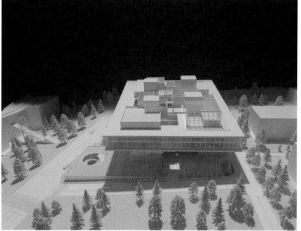

새로운 발상과 영감을 통해 무한한 관계와 협력을 바탕으로 창업과 교육 및 연구 등이 입체적으로 소통되고 결합되는 혁신을 유도하는 미래적인 가치에 열려진 용광로와 같은 공간이어야 한다. 포스코 기업을 넘어서 세계를 선도하고 미래에 대한 상상력을 펼치는 공간이며 새로운 유전자를 탄생시키는 인큐베이팅 장소다. 인큐베이팅 공간은 혁신의 아이콘을 가능케 하는 정신적인 가치를 담아내야 한다. 철을 녹여 쇳물을 만들고 철판과 소재를 만들어내는 용광로처럼 이 공간에는 영감, 네트워크, 커뮤니케이션의 가치를 담으며 서로 결합된 시너지를 창의적으로 발현시키는 역동적인 공간을 창출하여야 한다.

It should be a space like a furnace open to futuristic values that induce innovation in which entrepreneurship, education, and research are three-dimensionally connected and combined based on infinite relationships and cooperation through new ideas and inspiration. Leading the world beyond POSCO, it is a space of imagination open towards the future and an incubating place that creates new genes. The incubating space must contain the moral value that makes the icon of innovation possible. Like a furnace that melts iron to make molten iron and iron plates and materials, this space should create a dynamic space that contains the values of inspiration, network, and communication, and creatively expresses the combined synergy.

크리에티브 공유공간

창업지원센터는 서로의 작업을 공유하고 자극받으며 시너지를 만들어내는 창의적인 공간이 되어야 한다. 일반적인 연구소나 사무실 건축물이 각자가 연구하는 유니트공간에 집중되어 있다면, 포스코 체인지업 그라운드는 각각의 유니트 공간의 공유공간을 강화하여 가장 창의적인 소통을 만들어내려 하였다. 유니트공간보다 공유공간의 면적 비율을 최대한으로 만들어 주는 것을 기본으로 다양한 공유 프로그램을 재현하였다. 통상적으로 복도와 같은 공유공간을 확대하여 다양한 공간적 확장을 만들어낸다. 24시간 연구에 몰두하는 창업인들을 위한 리프레쉬공간이 되며, 다른 방식으로 새롭게 아이디어를 생성해내는 또다른 크리에티브 영역이 된다. 이 공유공간은 입체적인 포켓 정원의 개념을 함께 가지며 다양한 휴식의 질을 생성해낸다. 이 공유공간은 다양한 변형을 가진다. 폭과 높이의 다양한 공간적 변화를 통해서, 체육, 휴게, 자연, 숙면, 세탁 등 기본적인 창업을 위한 보완공간으로 변형되며 다양한 소통과 세미나를 위한 폴리형식, 오픈바, 정원, 세미나 공간 등으로 변형된다.

Creative Shared Space

The start-up support center should be a creative space where work is shared, stimulating one another, and synergy is created. While architecture for a research institute or office focuses on the unit space where an individual does research, POSCO aims to create the most creative communication by strengthening the shared space of each unit space. Various shared programs were reproduced based on maximizing the ratio of the area of the shared space rather than the unit space. In general, a shared space such as a hallway is expanded into a shared space to create various spatial extensions. It becomes a refreshing space for entrepreneurs who are immersed in 24-hour research, and it becomes another creative area to generate new ideas in a different way. This shared space has the concept of a three-dimensional pocket garden and creates variety in quality of relaxation. This shared space has various transformations. Through the variety of spatial changes in width and height, it is transformed into a complementary space for basic start-ups such as physical education, rest, nature, deep sleep, and laundry, and is transformed into a poly-type, open bar, and garden seminar space for various communication and seminars.

Public

공유형 연구공간

독립형 연구공간

유니버설 아트리움

전체를 관통하는 동선 장치이자 공유의 공간으로서 유니버설 아트리움을 구성한다. 하나로 구성된 입체적
아트리움 공간을 통해서 전체 레벨을 연결하고, 자연과 도시의 여러 프로그램과 공간이 자유롭게 결합된
복합적인 소통공간을 구성한다. 아트리움의 그라운드는 입체적인 단형의 로비공간으로 구성되어 다양한
이벤트를 가능케 하였다. 강의, 토론, 전시, 미디어, 영상 등을 통해서 창의적인 토론과 소통을 만들어 낸다.
소통의 세미나공간, 전시공간, 도시의 길, 숲, 휴식공간, 놀이공간 등이 입체적으로 구성된 무중력적인 공간이며
365일 조경이 살아있는 윈터가든의 광장과 같은 공간이다. 각층마다 아트리움이 면한 곳에 크리에이티브 박스를
다양하게 배치했다. 오픈된 아트리움 공간을 통해서 서로의 토론과 열정을 함께 공유한다. 그래서 보여지는
아트리움이 아니라 사용자들에게 열려지고 생활화되는 살아있는 공간이 된다.

Universal Atrium

It constitutes a universal atrium as a shared space and circulation device that penetrates the
whole. Through a three-dimensional atrium space composed as one, all levels are connected,
and various programs and spaces of nature and the city are freely combined to form a complex
communication space. The ground of the atrium is composed of a three-dimensional, short lobby
space, where different types of events can be held. Lectures, discussions, exhibition, media, and
videos create creative discussion and communication. It is a weightless space composed three-
dimensionally of seminar space for communication, exhibition space, urban roads, forest, lounge
space, play space, etc., and it is a space like a winter garden plaza where the landscaping is alive
365 days a year. On each floor, creative boxes were placed in various spots facing the atrium.
Discussions and passions are shared in the open atrium space. Therefore, it becomes not an
atrium for display, but a living space that is open to users and becomes a part of daily life.

PROCESS 01

공공성의 상아탑

Ivory Tower of Publicity

이 프로젝트는 대학 캠퍼스 내에 위치한 연구 시설로, 현대 사회에서 대학의 사회적 역할을 재고하게만 한다. 소위 '순수 학문을 지향'하는 대학은 재래적 상아탑의 감각에서 벗어나 내부에서 생성되는 창발적 사고와 연구의 사회적 기여에 대해 고민하여야 한다. 이 프로젝트는 학문과 연구의 성취가 고립된 담론으로 그치지 않고 사회와의 접점에서 그 가능성을 실험하는 공간이다.

This project is a research building located on a university campus, which causes the university reconsider its social role in modern society. Universities that pursue so-called "pure learning" must separate from the conventional ivory tower and think about the social contributions of emergent ideas and research generated from within. It is a space where the achievements of academics and research do not stop as isolated discourses, but experiment with their possibilities at the point of contact with society.

이러한 선행적 단계의 활동들은 '열린 상아탑'이며 '닫힌 광장'이라는 상반된 공간성을 필요로 한다. 내부로 침몰하는 숙고의 활동들이 일어나는 몰입의 공간은 발산의 통로로 유동적으로 전환될 수 있어야 한다. 이러한 동시성은 기존의 연구 공간과는 다른 새로운 건축적 유형(typology)을 요구한다. 도시화의 진행과 함께 전형화된 학교와 오피스의 공간들이 지향하는 내부적 완결성은 근대적 합리성이 체화(physicalisation)된 공간이다. 기계적 모듈 안에 균등하게 분배된 사무공간은 공동의 활동과 공간에 무관심한 태도를 보이며 그 이외의 공간 – 흔히 공용공간으로 분리되는 복도와 홀 등 – 은 이동과 점유에 있어 명확하게 용도를 확정하여 경제적으로 분배되어 있다. 경직된 공간은 그 곳에서 일어나는 활동들을 정확하게 지시하며 철저하게 관리될 수 있음을 전제한다.

그러나 이 프로젝트의 업무는 이러한 질서를 깨는 데서 출발한다. '책상 앞의 집중하는 활동'으로 관념화된 업무공간은 기계적으로 분류된 건축적 프로그램이며 그 공간에서 일어나는 다양한 양상의 활동들을 배제한다. 인간의 생산 활동을 기계적 작동으로 가정한 이러한 공간의 프로그래밍은 몰입과 휴식, 공유 등의 다양한 양상의 활동들이 '일'의 영역으로 확장되어 '업무공간'이라는 프로그램에 근거하여 실을 구획하는 방식은 인간의 생산 활동을 기계적 작동으로 가정한다. 그러나 이 프로젝트는 명명되지 못하는 혹은 이를 거부하는 다양한 정체성의 공간들을 통해 업무와 연구의 다양한 모습들을 담는다.

몰입과 휴식 그리고 공유 등의 다양한 '활동의 스펙트럼(activity spectrum)' 과 이러한 활동들의 농도를 조절하는 디자인 방법을 통해 확정적이지 않은 공간에서 다양한 활동(activity)들이 혼재되고 복합된 형태로 나타난다. 경직된 공간이 활동의 느슨함으로 간섭되고 분해되어 빈틈을 만들어 내고 가능성의 공간으로 비워진 곳은 연구자들의 내부적 광장으로 작동한다. 이러한 광장은 연구의 활동들이 사회와의 접점을 생성하기 이전의 전이공간적 성격을 가진다. 사회와의 접점은 단일한 통로가 아닌 다양한 공간적 요소들을 갖춘 입체적 표피(spatial skin)로 연구와 창업, 개별공간과 공유공간, 대학과 지역 사회 등 상이한 요소들간에 다양한 관계성을 제시한다. 학문의 공간이 공동체적 가치를 공유하며 내부적 생태계를 형성하는 관계들을 발현시킬 수 있는 공동의 장으로 확장될 때 교육기관의 능동적 사회적 역할을 다할 수 있을 것이다.

These activities are "open ivory towers" and require an opposite spatiality of "closed squares". The space of immersion, where the activities of contemplation sink into the inside, must be able to be fluidly converted into a channel of divergence. This simultaneity requires a new architectural typology that is different from the existing research space. With the progress of urbanization, the interior completion of the stereotypical school or office is a space in which modern rationality is embodied (physicalization). Office space evenly distributed in the mechanical module shows an indifferent attitude to common activities, and other places – such as corridors and halls that are often divided into common areas – that are economically distributed by clearly determining their use for movement and occupation. A rigid space presupposes that the activities taking place there can be precisely and thoroughly managed.

However, the work of this project begins with disregarding this order. The work space conceptualized as "concentrated activity in front of a desk" is a mechanically classified architectural program that excludes various types of activities occurring in the space. The programming of this space, assuming human production activity is a mechanical operation, expands various activities such as immersion, rest, and sharing into the field of "work," and the method of dividing rooms based on the program "workspace" assumes that human production activities are mechanical operations. However, this project contains aspects of work and research through spaces of various identities that are denied or cannot be named.

Numerous activities appear in a mixed and complex form in an indeterminate space through different "activity spectrums" such as immersion, rest, and sharing, and a design method controls the concentration of these activities. The rigid space is interfered with and decomposed by the looseness of activity to create gaps, and the area emptied as a place of possibility functions as an inner square for researchers. The square has the characteristics of a transitional space before research activities create a contact point with society.

The point of contact with society is not a single passage, but a three-dimensional spatial skin equipped with assorted spatial elements, suggesting a mixture of relationships between different components such as research and startup, individual and shared spaces, and universities and local communities. When the academic space is expanded into a communal field that shares community values and expresses the relationships that form an internal ecosystem, the educational institution will be able to fulfill its active social role.

PROCESS 02

계열체 간의 관계성을 위한 구축 실험들
Construction Experiments for Relationships Between Succession Systems

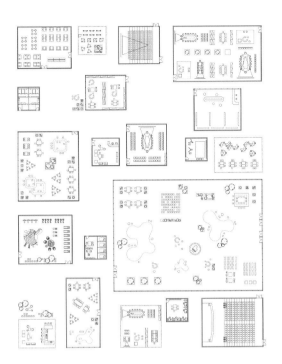
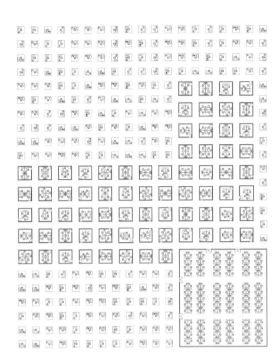

건축을 하나의 완성된 글로 치환하여 생각해보면, 건축의 기본 요소들 ‒ 기둥, 벽, 바닥, 천장 등 ‒ 은 홀로는 의미가 없는 알파벳에 비유가 될 수 있다. 그리고 이러한 요소들이 서로에게 의지하여 만들어진 공간적 요소는 비로소 특정 의미를 생성하기 시작하는 단어, 또는 구절에 비유될 수 있다. 이러한 조합물은 특정 언어법칙으로 구조화될 때 의미를 생성하며 완결성을 획득하고 틀 안에 대체 가능한 다양한 계열체들로 확장성을 얻는다. 완결성과 확장성은 대비되는 특질로 보이지만 관계하며 의미를 생성하는 부분과 전체의 구조를 이루는 유연함 속에서 이루어질 수 있다. 이 프로젝트는 대체 가능한 계열체들의 조합으로 전체를 구성하는 건축적 방법론으로 다양한 구축의 실험들을 통해 완성되었다.

If we think of architecture as a complete text, the basic elements of architecture – pillars, walls, floors, ceilings, etc. – can be compared to an alphabet that has no meaning by itself. And the spatial elements created by these elements can be compared to words or phrases that begin to create specific meanings. When these combinations are structured with specific linguistic laws, they generate meaning, acquire completeness, and expandability is achieved by having various substitutable succession systems within this framework. Although completeness and expandability appear to be contrasting qualities, they can be achieved in the flexibility that constitutes the structure of the part and the whole that are related and generate meaning. This project was completed through several experiments in construction with an architectural methodology that unites the whole with a combination of succession subsystems.

이 프로젝트는 공간으로서의 최소 의미를 가지는 큐브(cube)에서 시작한다. 6개의 면으로 구성된 큐브는 벽과 바닥의 실체를 지우면서, 독립적 공간으로 작동하는 기본 요소로 정의된다. 이는 종래의 건축적 기본 요소들 – 기둥, 벽, 바닥, 천장 – 은 소거되어 '방 房 : room'을 만드는 구축을 부정하는데서 출발한다. 건축적 질서로 분절되기 이전의 단위 공간인 큐브에 대한 의미는 잠재적인 부재 상태를 유지하며 관계를 전제로 존재한다. 이렇게 느슨한 공간적 테두리 이상의 의미를 갖지 못하는 큐브들은 그 곳에 위치할 필연성, 혹은 당위성을 가지지 않으며 이들의 집합은 건축가의 유형의 틀이 생성되기 이전부터 존재하는 상호간에 대체 가능한 계열체다. 복수의 큐브들이 공간적 정체성을 발현하기 시작하는 것은 이들의 관계 맺기를 통해서이다. '이 곳에 있어야만 하는: must be' 확정적 지위를 갖지 않는 큐브들의 관계성을 정의하고 그에 따른 결합 방식에 따라 다양한 양상의 통합체를 만든다. '큐브'의 기본형은 몰입을 위한 개인적 공간이며 내부적 연구 생태계를 생성하기 위한 공유적 연구공간과 사회와의 접점을 확대하기 위한 지원 공간들에 의해 공공성의 상아탑으로 확장된다. 큐브들의 관계성의 물화는 건축의 형태로 직접적으로 발현되며 공유공간의 개념은 건축으로 치환됨에 있어 직접적이고 선명한 태도를 유지하며 조형적으로 발현되는데 있어 그 적나라함을 검열하지 않는다.

The project starts with a cube that has minimal meaning as a space. The six-sided cube is defined as a basic element that operates as an independent area while erasing the reality of walls and floors. This begins denying the construction that creates a "room" by erasing the basic architectural elements- columns, walls, floors, and ceilings. The meaning of the cube, the unit space before being segmented into architectural order, maintains its potential absence and exists on the premise of a relationship. Cubes, which have no meaning beyond these loose spatial boundaries, do not have the necessity or justification to have a specific location, and their set is a mutually interchangeable sequence that exists before the architect's type frame was created. It is through their relationships that a group of cubes begins to express their spatial identity. "Must be here" defines the relationships of cubes that do not have a definitive status, and creates various types of unity according to the combination method. The basic form of a "cube" is a personal space for immersion, and it is expanded into an ivory tower of publicity by a shared research place in which to create an internal research ecosystem and support spaces to expand contact points with society. The reification of the relationships of the cubes is expressed directly in the form of architecture, and the concept of shared space maintains a direct and clear attitude when it is replaced by architecture, and does not censor its nakedness in its formative expression.

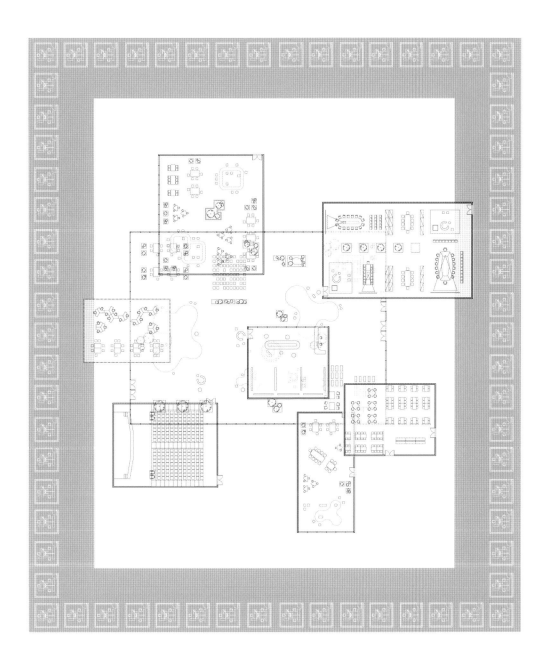

Penetration : 관입 / 해체를 위한 구축

개별적 연구 공간, 몰입의 공간들은 장방형의 매스로 구축된다. 이는 프로그램의 분류와 분배의 측면에서 명확함과 경제성을 보여주는 근대 건축이 성취한 공간적 논리를 유지한다. 가장 평이한 방법으로 구축되는 전형들, 개별 공간의 수평적 배치와 그 구성의 규칙 안에서 조형적 리듬을 구성하는 작업이 선행된다. 여기서 이러한 근대적 건축 논리의 물화는 다분히 의도적인 작업으로 새로운 구축을 위한 과거의 구축을 부정하기 위한 선행 단계이다. 개별 연구 공간들이 일렬로 배치되어 구성된 견고한 장방형의 매스는 해체를 위한 구축이며, 분해를 위한 조합이다. 이러한 몰입을 위한 개인적 공간들의 집합은 대지의 북동 측과 남서 측 끝에 두 동으로 배치되고 그 사이를 관입하는 매스들은 전형적 연구실의 공간 구성을 해체하며 이형의 공간들을 만들어 낸다. 직교 체계의 질서를 부정하며 관입된 공간들은 형태적 경직성을 깨는 것과 동시에 공유의 프로그램들의 무작위적으로 발생하는 지대를 만든다. 또한 대지의 중앙부를 비우면서 생겨난 수직적 틈(Void)을 통해 캠퍼스의 경관 축이 생성되며 이곳에는 연구와 실험이 외부적으로 확장되는 다양한 공유 활동들의 풍경으로 채워진다.

Penetration: Building for Intrusion / Dismantling

Individual research spaces and areas of immersion are constructed with a rectangular mass. This maintains the spatial logic achieved by modern architecture, which shows clarity and economy in terms of program classification and distribution. The work of composing the formative rhythm within the patterns constructed in the simplest way, the horizontal arrangement of individual spaces and the rules of its composition, comes earlier. Here, the reification of this modern architectural logic is a deliberate work, and it is a prior step to deny the construction of the past for a new construction. The individual study spaces are arranged in a row and the solid rectangular mass is a construction and a combination for disassembly. These sets of personal spaces for immersion are arranged in two buildings at the northeast and southwest ends of the site, and the masses erected between them deconstruct the spatial composition of a typical laboratory and create heterogeneous intervals. The intrusive spaces, which negate the order of the orthogonal system, break the rigidity of form and at the same time create a zone of random occurrences of shared programs. In addition, the landscape axis of the campus is established through a vertical void created by emptying the central part of the site, and filling it with a scene of various shared activities where research and experimentation are expanded outwardly.

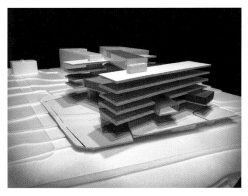
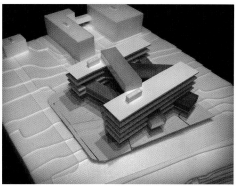

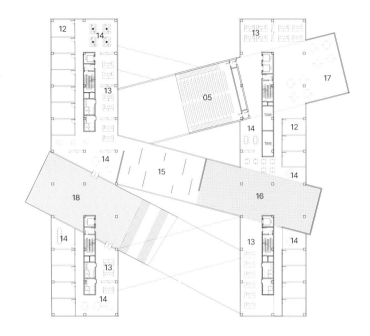

3rd Floor Plan

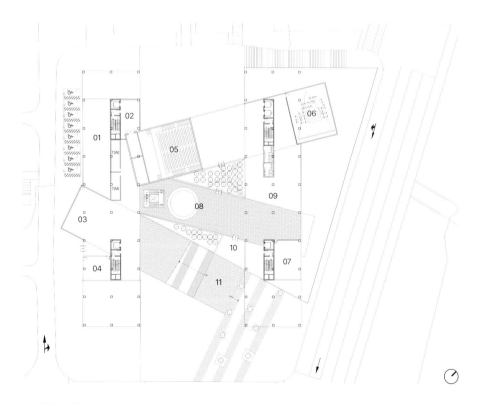

2nd Floor Plan

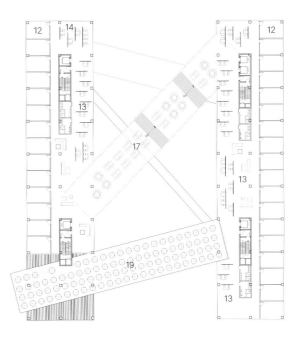

5th Floor Plan

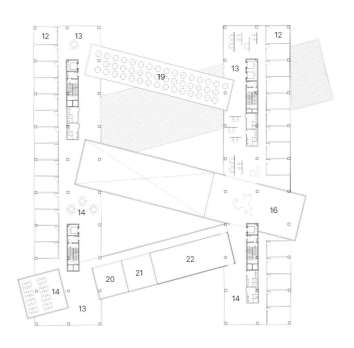

4th Floor Plan

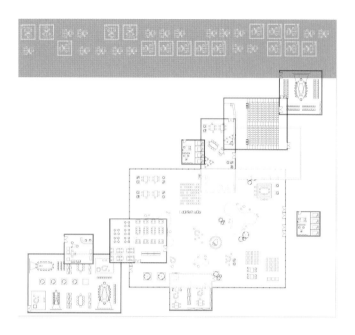

Integration : 통합 / 병렬적 연결의 해체

회랑 형식으로 둘러싼 개별 연구실들은 큐브의 병렬적 결합으로 각각의 큐브는 독립된 완결성을 가진다. 개별 큐브들은 더 이상 결합되지 않고 그저 통합을 지향하는 파편으로 서로 병치되어 존재한다. 지상에서 띄워진 연속된 큐브 회랑은 도시와 접점을 갖지 않는 의도된 고립성을 획득한다. 이러한 큐브의 병렬적 질서는 건물의 중앙부에서 흩어지고 병합되며 재배열되어 지식과 연구의 활동들이 다양한 형태로 발산되며 공유되는 공간을 만든다. 이러한 중앙부의 공동의 프로그램들은 수직적으로 관통하는 아트리움을 통해 지상과의 접점을 구성하며 지역 사회와 소통하는 공간을 조성한다. 전시 세미나 등을 통해 연구자 간의 소통과 협력의 공유 공간은 더 나아가 사회 속의 살아 있는 상아탑의 공간으로 확장된다.

Integration : Breakdown of Parallel Connections

Individual labs opening off a corridor are parallel combinations of cubes, so each cube has an independent completeness. Individual cubes are no longer combined, but exist alongside each other as fragments oriented toward integration. The corridor of continuous cubes floating from the ground acquires an intended isolation that does not have contact with the city. The parallel order of these cubes diverges, merges, and rearranges in the central part of the building, creating a space where knowledge and research activities are radiated and shared in various forms. These common programs in the central part of the project form a contact point with the ground through a vertically penetrating atrium and create a space in which to communicate with the local community. Through exhibitions and seminars, the shared space of communication and cooperation among researchers is further expanded into the space of an ivory tower surviving in society.

01 Meeting Room

02 Cooperative Space

03 Occupancy Space(Independent Type)

04 Occupancy Space(Open Type)

05 Small Library

06 Communication Zone

07 Event Hall(Large)

08 Cafe

09 Storage

10 Parking

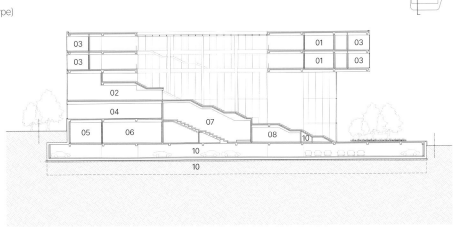

Section

PROCESS 03

무용한 공간의 유용성
Usability of Useless Space

앞서 얘기한 세 가지의 디자인은 모두 연면적의 상당 부분이 공용공간으로 이루어져 있다. 공용공간은 모든 종류의 건축에 등장한다. 흔히, 복도, 계단과 같은 수평, 수직적 이동과 화장실, 로비 등과 같이 사용자 모두의 기본적 편의를 위한 공간이기 때문에 공용공간은 '필요한' 공간이지만 건축이 목표로 하는 특정 프로그램을 수용하는 그 외의 공간이므로 최소한의 면적으로 계획되어야 합리적인 디자인으로 간주되어 왔다. 기계적 장치로서의 건축은 필요한 공간은 명확하게 범주화(zoning) 되고 이러한 건축적 프로그램들은 경제적으로 잘 작동될 있도록 계획되는데 있어 공용공간은 합리성 그 이상의 지위를 갖지 않는다. 이러한 공간적 위계는 경제성과 합리성을 이상적 가치로 여기던 근대적 사고에 건축이 일조해 왔던 결과물일 것이다. 그러나 공간을 점유하는 인간은 특정적 한가지 활동만을 하도록 제어되는 공간의 목적성을 배반한다. 완벽하게 편안한 소파가 있음에도 화장실에서 집중이 더 잘되는 경험, 주방에서 식탁으로 경제적 동선에도 불구하고 모니터 앞에서 무릎을 세운 채 먹는 식사와 같은 경험은 공간이 특정하는 프로그램과 그에 따른 명확성의 건축이 허상임을 보여준다. 건축가가 제시하는 공간의 프로그램과 사용자의 활동 사이의 간극에서 정체성이 부재한 공간의 필요성이 자각된다. '사무실에서는 업무', '복도에서는 이동'과 같이 계획된 건축 프로그램에 따라 분리된 공간에서도 사용자의 점유 방식은 그 기대를 벗어나 이루어진다. 업무에 몰두한 연구자는 일어나서 잠시 숨을 고르는 시간에도 사고의 확장을 경험하며 말없이 집중할 동료를 보는 순간, 같은 공간 속에서 일어나는 무언의 공동체성을 느낀다. 이러한 모호한 활동들이 일어나며, 혹은 일어나지 않아도 괜찮은 일종의 무용한 공간이 허락되어야 한다. 이러한 공간의 조성에 있어서 중요한 요소는 변칙성이다. 직선형의 공용공간은 건축공간이 존재하는 체계를 명확하게 보여줌과 동시에 그 속에서 사용자는 공간의 점유 방식에 대한 방향성을 암시하고 있다. 건축화된 프로그램들간의 관계의 선이자 시스템을 결정짓는 회로도의 경직성을 보여준다. 공유하는 공간이 아닌 그 안에서의 활동이 경계없이 흘러나올수 있는 공간이어야 한다. 건축의 질은, 공간성의 다양함은 공용공간의 다양함에서 얻을 수 있다.

확정된 건축의 프로그램은 건축의 공간적 질을 결정한다. 이 프로젝트의 경우, 업무 공간의 질이 건축의 성취도를 결정하며 이는 기능적으로 충족되는 충분 조건인 수동적 디자인이다. 도시의 역동성은 거리에서 일어나는 커뮤니티의 양과 질에 비례하며 이는 건축 내부의 생태계에서도 유효하다. 동적인 질서는 근대적 의미의 이동을 위한 단일한 의미에서 불확정적 이벤트가 일어날 수 있는 잠재성의 공간으로 확장된다.

In all three designs mentioned above, a significant part of the total floor area is made up of common spaces. Common spaces appear in all types of architecture. Common space is a "necessary" space because it is for the basic convenience of all users, with horizontal and vertical movement such as corridors and stairs and restrooms. It is considered a reasonable design when it is planned with a minimum area, as it is a space other than one that accommodates a specific program targeted by architecture. In architecture, as a mechanical device, the necessary spaces are clearly zoned and these architectural programs are planned to work well economically, and the public space has no status beyond rationality. This spatial hierarchy is probably the result of architecture's contribution to modern thinking, which regarded economic efficiency and rationality as ideal values.

However, the person occupying the space betrays the purpose of the space, which was structured to do only one specific activity. Experiences such as having a better concentration in the bathroom even with a perfectly comfortable sofa in the space, and eating off of one's knees in front of a monitor despite the economic flow from the kitchen to the table, show that the space-specific program and the resulting architecture of clarity are an illusion. In the gap between the spaces in the program presented by the architect and the user's activities, the need for a space without identity is recognized. Even in an area separated according to the architectural program, such as work in an office and movement in the hallway, the user's occupancy method goes beyond expectations.

Researchers who are immersed in their work experience an expansion of their thinking, even when they get up to catch their breath, and feel the unspoken sense of community that occurs in the same space even as they silently watch the concentration of their colleagues. A kind of useless space must be allowed for these vague activities to take place or not to happen. An important factor in the composition of these area is anomaly. The rectilinear common space clearly shows the system in which the architectural space exists, and at the same time, the user hints at the direction of occupying the space. It shows the rigidity of the circuit diagram that determines the relationships between the built programs and the system. It should not be a shared space, but a space in which activities can flow without boundaries. The diversity of spatiality and quality of architecture can be obtained from the mixture of public spaces.

The established architectural program determines the spatial quality of the architecture. For this project, the quality of the work space determines the achievement of the architecture, which is a passive design where functional fulfillment is sufficient. The dynamism of the city is proportional to the quantity and quality of the community that takes place on the street, and this is also valid in the environment inside the building. The dynamic order extends from a single to a space of potential where spontaneous events can occur.

BUILDING

Site Plan | Section | Plan | Photo

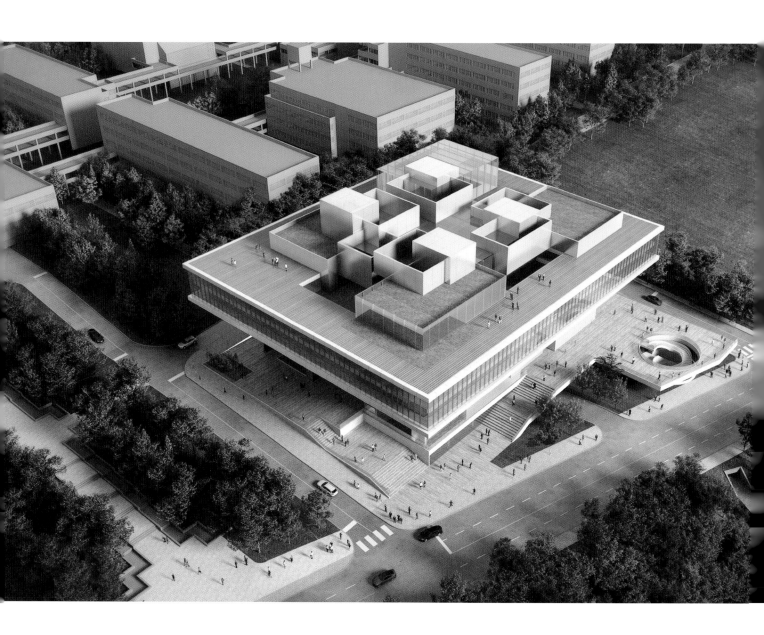

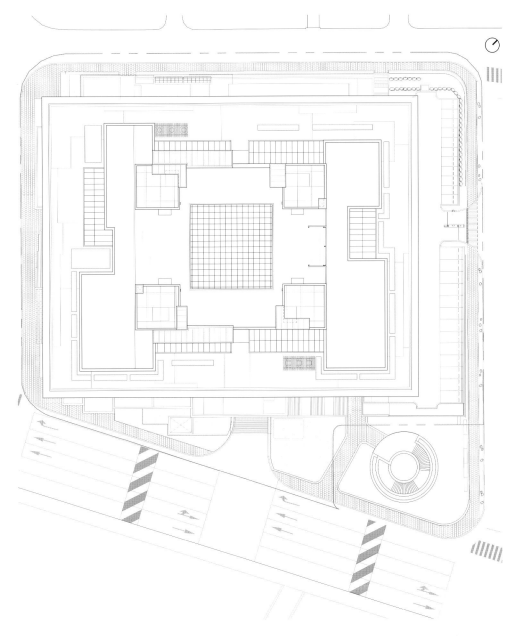

Site Plan

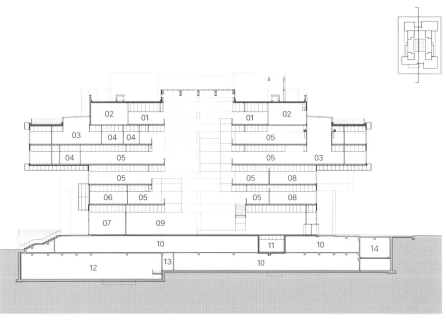

| 01 Hall |
| 02 WET LAB |
| 03 Terrace |
| 04 Meeting Room |
| 05 Lounge |
| 06 Education / Working Space |
| 07 Neighbourhood Facility |
| 08 Office |
| 09 Lobby |
| 10 Parking |

Section

11 Sports Room

12 Electrical Room

13 Storage

14 Ramp

15 Studio

16 Play Zone

17 Start-Up Culture Space

18 Server Room

19 Generator Room

20 Mechanical Room

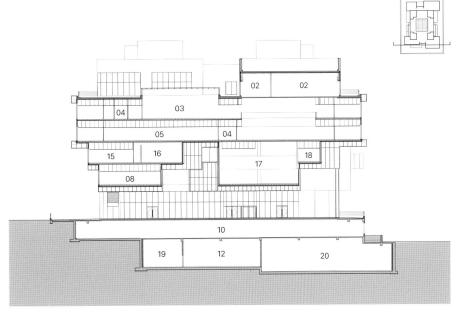

Section

3rd Floor Plan

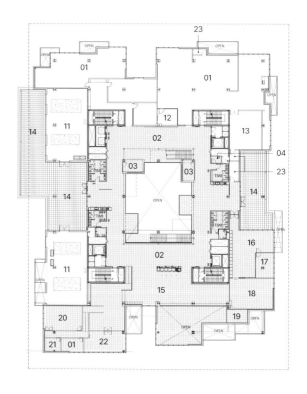

4th Floor Plan

01 Office

02 Lounge

03 Meeting Room

04 Cleaning Tool Room

05 Pantry

06 Start-up / Working Space

07 Education / Working Space

08 Start-up Culture Space

09 Lecture Room

10 Control Room

11 Air Handling Unit Room

12 Conference Room

13 Main Conference Room

14 Terrace

15 Multi-Purpose Room

16 Administration Office & Lab

17 Director's Office

18 Education Space

19 Server Room

20 Studio

21 Video Production Room

22 Play Zone

23 Storage

24 Locker Room

25 Laundry Room

26 Rest Room

27 Hall

28 Shower Room

29 Corridor

30 Phone Booth

31 WET LAB

32

33 Hall

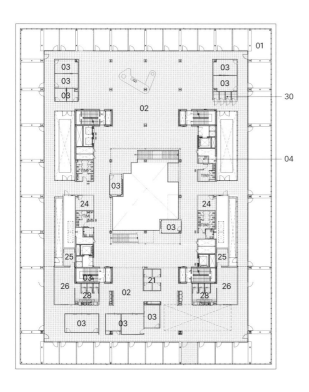

5th Floor Plan

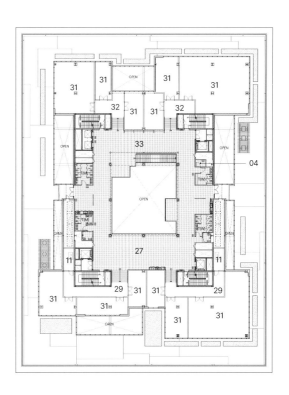

7th Floor Plan

EPILOGUE

어크로스 더 유니버스
Across the Universe

  기차역에서 내려 택시를 타 일단 포항공과대학교로 가달라고 했다. 캠퍼스 안 적당한 곳에 내려 찾아보면 되겠지 했는데, 캠퍼스 안으로 진입하자마자 찾을 것도 없이 기세 등등하게 모습을 드러낸다. 체인지업그라운드 포항은 캠퍼스 주진입도로 변에 중앙도서관과 마주한 자리에 위치한다. 북쪽의 강의동들과 남쪽으로 연구동들 사이에 위치하며, 도로로 나뉘어진 동서축을 연결하는 키스톤 같은 역할을 한다. 가로 60m, 세로 80m에 이르는 두 개 층 높이의 큰 처마 같은 판상형 매스에 금속 소재로 마감한 육면체들이 매달려 있는 첫인상은 우주를 가로질러 캠퍼스에 불시착한 우주선처럼 이질적이면서 동시에 미래적인 느낌이다. 판상형 매스의 캔틸레버가 좀 더 길게 빠져 아래에 매달린 덩어리들과 대비가 좀 더 극명했으면 좋았겠다고 지나가듯 한 얘기에 실시설계 과정에서 내진이슈와 공사비 문제로 조정이 되었다고 건축가는 길게 아쉬워한다.

  내부로 들어가니 6개층 높이의 아트리움과 그 주변을 둘러싼 육면체 방들의 조합, 그리고 각 층을 연결하는 선형계단의 흐름이 먼저 눈에 들어온다. 외부 이미지가 내부의 공간구성으로 연결되어 유기적이고 통합적 환경을 구축하고 있다는 인상이다. "21세기 세계는 평평하다"는 토마스 프리드만의 주장처럼 역사적·지리적 분리가 점점 더 무의미해져 가는 무한경쟁의 시대이고 경쟁력을 유지하기 위해 기업과 개인들에게 더욱 많은 개인적·집단적 창의력과 협업이 요구된다. 따라서 최근 업무공간에서는 단순히 효율만을 고려하는 것이 아니라 몰입과 소통, 휴식 공간의 균형을 맞추면서 상호간 쉽게 모드가 전환될 수 있는 복합적 성격의 공간 계획이 요구되고 있다. 설계 의뢰를 받을 때 여러 기능 공간에 대한 요구가 있었을 것이다. 이를 어떻게 분류하고 조합하며 공간별로 어떤 특성을 부여할 것인지를 정하는 것에서 건축가의 해석과 철학이 개입된다. 건축가는 이 인큐베이팅 센터의 정체성을 드러내는 공간, 몰입 공간으로서 개별연구공간, 협업을 위한 공유적 연구 공간, 그리고 휴식과 소통의 공간 이렇게 네 개의 카테고리로 공간을 분류하고, 이 공간 간의 관계를 설정하는 시스템이 곧 건축적 형태로 드러나도록 하고 싶었다고 한다. 다양한 공유공간들이 자유롭게 결합하여 채워진 아트리움과 그 주변을 에워싸고 있는 개별 또는 공유의 연구공간들, 중간중간 자연채광이 되는 중정과 결합된 휴식공간들이 입체적으로 엮이면서 만들어내는 공간감은 조화롭게 잘 작동하는 생태계를 시각적으로 구현해 놓은 것 같다.

체인지업그라운드 포항은 언뜻 봐도 공유적 성격으로 열려있는 공간의 면적비율이 전체면적의 절반 가까이 되어 보이는 개방적 성격의 내부공간 구성을 갖는다. 이는 공용공간이 기능적 실들을 연결하는 이동공간으로서 존재하던 지난 세기에 지어진 건물들과 큰 차이를 보인다. 건물과 건물 사이의 경계에 위치한 모호한 성격의 옥외공간(공공공간 또는 중간영역)에서 예기치 못한 다양한 커뮤니티 활동이 일어난다는 얀 겔의 이론이 내부공간에서 효과적으로 입증된 느낌이다.

장윤규는 "운생동의 작업방식은 가능한 모든 대안을 만들고 거기서부터 추려나가는 귀납적 방식으로, 지난 십 여 년에 걸친 작업의 과정에서 중간결과물로 생산된 건축적 아이디어들이 상당히 축적되어 있다"고 했다. (이 대목에서 엄청 부러웠다.) 이런 작업방식은 특정 프로그램에 딱 맞는 맞춤옷을 만들어 내는 건축생산방식과는 달리 다양한 조건에 대응 가능한 확장성을 가질 수 있는 방식일 수 있다. 또한 운생동의 프로젝트들은 단위 유니트의 조합으로 전체를 만들어내는 전략을 많이 취하는 데, 이는 형태적으로나 내부공간구성에 있어서 복합성과 생동감을 구현할 수 있는 영리한 전략이다. '(기)운생동'이라는 회사명의 의미와 같이 운생동의 건축에서는 동적인 에너지가 느껴진다. 대학(university) 캠퍼스 내에 하나의 우주(universe)와 같은 생태계를 구현한 포스코 인큐베이팅 센터는 비단 창업지원센터로서만이 아니라 전유공간과 공유공간의 구축 논리에 있어서 이 시대에 맞는 하나의 전형이 될 수 있다. 코로나 이후 이 기운생동의 공간에서 다양한 이벤트가 일어나고 사람들이 분주하게 오가고 치열하게 연구하고 곳곳에서 생각을 나누고 협업하는 광경을 보게 될 날이 기다려진다.

I got off a train and took a taxi to go to Pohang University of Science and Technology. I thought I could get down to a suitable place on campus and find what I was looking for, but as soon as I entered the campus, I couldn't find anything. The CHANGeUP GROUND is located on the side of the main entrance road to the campus and faces the Central Library. It is located between the lecture buildings in the north and the research buildings in the south, and serves as a keystone connecting the east-west axis divided by roads. The first impression of metal cubes are suspended from a large eaves-like mass that is 60m wide, 80m long and 2 stories high gives a foreign and futuristic feeling like a spaceship crash-landed on campus. The architect regretted for a long time that it was altered during the detailed design process because of seismic resistance and construction cost issues, saying it would have been better if the cantilever of the plate-shaped mass was a little longer and the contrast with the protuberances hanging below was sharper.

Upon entering the interior, the first thing that catches the eye is the combination of a six-story atrium, the hexagonal rooms surrounding it, and the flow of the linear staircase connecting each floor. It gives the impression that the external image is connected to the internal spatial composition to construct an organic and integrated environment. As Thomas Friedman asserts, "the world in the 21st century is flat," It is an era of unlimited competition in which historical and geographic separation is increasingly meaningless, and more individual, and collective creativity, and collaboration are required from companies and individuals to maintain competitiveness. Therefore, in recent work spaces, there is a demand for a space plan with a complex nature that can easily switch modes while balancing immersion, communication, and relaxation spaces, rather than simply considering efficiency. When receiving a design request, there must have been a request for several functional spaces. The architect's interpretation and philosophy are involved in determining how to classify and combine them, and what characteristics to give to each space. The architect classifies the space into four categories: an area that reveals the identity of the incubating center, an individual research place as an immersion space, a shared research room for collaboration, and a space for relaxation and communication. It is said he wanted the system to establish the relationships between these spaces to be revealed in an architectural form.

The atrium is filled with various shared spaces freely combined, with individual or shared research room surrounding it, and the relaxation areas combined with the courtyard providing natural light in the middle are three-dimensionally interwoven to create a harmonious sense of space. It seems to be a visual application of the ecosystem.

At first glance, the CHANGeUP GROUND has an open interior configuration, where the area ratio of open spaces is close to half of the total area due to its shared nature. This is very different from the buildings built in the last century, where the common space existed as an area connecting functional rooms. It appears that Jan Gel's theory that various spontaneous community activities occur in an outdoor space (public space or intermediate area) of an ambiguous nature located at the boundary between buildings is effectively proven in the interior space.

Yoon-gyu Jang said, "Unsaengong's work method is an inductive approach of generating all possible alternatives and selecting from them. Architectural ideas produced as intermediate results in the process of work over the past ten years have accumulated considerably." (I was very envious at this point.) This work method can be scalable to respond to various conditions, unlike the architectural production method that creates "tailored clothes" that are perfect for a specific program. In addition, many projects by Unsaengdong Architects follow a strategy of creating a whole through the combination of single units, which is a clever strategy for realizing complexity and vitality in form and internal space composition. As in the meaning of the company name "(Gi) Unsaengdong," the architecture of Unsaengdong projects dynamic energy. The POSCO Incubating Center, which contains an ecosystem reflecting a universe within the university campus, can be a model suitable for this era not only as a start-up support center, but also in the logic of building exclusive and shared spaces. I look forward to the day when we will see various events taking place in this lively building after Covid-19, people busily coming and going, doing intense research, sharing thoughts and collaborating everywhere.

체인지업그라운드 포항

건축사사무소	포스코 A&C + 운생동건축사사무소
건축가	장윤규
글쓴이	장윤규
	김미정
	김정임
사진	남궁선
제작	마실와이드
ISBN	978-89-968146-5-8(94540)
	978-89-968146-4-1 (세트)

Architecture	POSCO A&C + UNSANGDONG Architects
Architect	Jang Yoon Gyoo
writer	Jang Yoon Gyoo
	Kim Mi Jung
	Kim Jeong Im
Photography	Namgoong Sun
Production	MasilWIDE
ISBN	978-89-968146-5-8(94540)
	978-89-968146-4-1 (SET)